Joint FAO/WHO Food Standards Programme
CODEX ALIMENTARIUS COMMISSION

CODEX ALIMENTARIUS

FOOD HYGIENE
BASIC TEXTS

THIRD EDITION
2003

FOOD AND AGRICULTURE ORGANIZATION
OF THE UNITED NATIONS
WORLD HEALTH ORGANIZATION

ISBN 92-5-105106-2

PREFACE

THE CODEX ALIMENTARIUS COMMISSION AND THE FAO/WHO FOOD STANDARDS PROGRAMME

The Codex Alimentarius Commission implements the Joint FAO/WHO Food Standards Programme, the purpose of which is to protect the health of consumers and to ensure fair practices in the food trade. The *Codex Alimentarius* (Latin, meaning Food Law or Code) is a collection of internationally adopted food standards presented in a uniform manner. It also includes provisions of an advisory nature in the form of codes of practice, guidelines and other recommended measures to assist in achieving the purposes of the Codex Alimentarius. The Commission has expressed the view that codes of practice might provide useful checklists of requirements for national food control or enforcement authorities. The publication of the Codex Alimentarius is intended to guide and promote the elaboration and establishment of definitions and requirements for foods, to assist in their harmonization and, in doing so, to facilitate international trade.

BASIC TEXTS ON FOOD HYGIENE – THIRD EDITION

The basic texts on food hygiene were adopted by the Codex Alimentarius Commission in 1997 and 1999. This is the third edition of this compact booklet first published in 1997 and includes the revised Guidelines for the Application of the HACCP System adopted by the Codex Alimentarius Commission in 2003. It is hoped that this compact format will allow wide use and understanding of the basic principles of food hygiene and that it will encourage their use by governments, regulatory authorities, food industries and all food handlers, and consumers.

Further information on these texts, or any other aspect of the Codex Alimentarius Commission, may be obtained from:

> *The Secretary, Codex Alimentarius Commission,*
> *Joint FAO/WHO Food Standards Programme,*
> *FAO, Viale delle Terme di Caracalla,*
> *00100, Rome Italy*
>
> *fax: +39(6)57.05.45.93*
> *email: codex@fao.org*

CONTENTS

RECOMMENDED INTERNATIONAL CODE OF PRACTICE
GENERAL PRINCIPLES OF FOOD HYGIENE

CAC/RCP 1-1969, Rev. 4 (2003)

INTRODUCTION

People have the right to expect the food they eat to be safe and suitable for consumption. Foodborne illness and foodborne injury are at best unpleasant; at worst, they can be fatal. But there are also other consequences. Outbreaks of foodborne illness can damage trade and tourism, and lead to loss of earnings, unemployment and litigation. Food spoilage is wasteful, costly and can adversely affect trade and consumer confidence.

International food trade, and foreign travel, are increasing, bringing important social and economic benefits. But this also makes the spread of illness around the world easier. Eating habits too, have undergone major change in many countries over the last two decades and new food production, preparation and distribution techniques have developed to reflect this. Effective hygiene control, therefore, is vital to avoid the adverse human health and economic consequences of foodborne illness, foodborne injury, and food spoilage. Everyone, including farmers and growers, manufacturers and processors, food handlers and consumers, has a responsibility to assure that food is safe and suitable for consumption.

These General Principles lay a firm foundation for ensuring food hygiene and should be used in conjunction with each specific code of hygienic practice, where appropriate, and the guidelines on microbiological criteria. The document follows the food chain from primary production through to final consumption, highlighting the key hygiene controls at each stage. It recommends a HACCP-based approach wherever possible to enhance food safety as described in *Hazard Analysis and Critical Control Point (HACCP) System and Guidelines for its Application* (Annex).

The controls described in this General Principles document are internationally recognized as essential to ensure the safety and suitability of food for consumption. The General Principles are commended to Governments, industry (including individual primary producers, manufacturers, processors, food service operators and retailers) and consumers alike.

SECTION I - OBJECTIVES

THE CODEX GENERAL PRINCIPLES OF FOOD HYGIENE:

- identify the *essential* principles of food hygiene applicable *throughout the food chain* (including primary production through to the final consumer), to achieve the goal of ensuring that food is safe and suitable for human consumption;

- recommend a HACCP-based approach as a means to enhance food safety;

- indicate *how* to implement those principles; and

- provide a *guidance* for specific codes which may be needed for - sectors of the food chain; processes; or commodities; to amplify the hygiene requirements specific to those areas.

SECTION II - SCOPE, USE AND DEFINITION

2.1 SCOPE

2.1.1 THE FOOD CHAIN

This document follows the food chain from primary production to the final consumer, setting out the necessary hygiene conditions for producing food which is safe and suitable for consumption. The document provides a base-line structure for other, more specific, codes applicable to particular sectors. Such specific codes and guidelines should be read in conjunction with this document and *Hazard Analysis and Critical Control Point (HACCP) System and Guidelines for its Application* (Annex).

2.1.2 ROLES OF GOVERNMENTS, INDUSTRY, AND CONSUMERS

Governments can consider the contents of this document and decide how best they should encourage the implementation of these general principles to:

- protect consumers adequately from illness or injury caused by food; policies need to consider the vulnerability of the population, or of different groups within the population;

- provide assurance that food is suitable for human consumption;
- maintain confidence in internationally traded food; and
- provide health education programmes which effectively communicate the principles of food hygiene to industry and consumers.

Industry should apply the hygienic practices set out in this document to:

- provide food which is safe and suitable for consumption;
- ensure that consumers have clear and easily-understood information, by way of labelling and other appropriate means, to enable them to protect their food from contamination and growth/survival of foodborne pathogens by storing, handling and preparing it correctly; and
- maintain confidence in internationally traded food.

Consumers should recognize their role by following relevant instructions and applying appropriate food hygiene measures.

2.2 USE

Each section in this document states both the objectives to be achieved and the rationale behind those objectives in terms of the safety and suitability of food.

Section III covers primary production and associated procedures. Although hygiene practices may differ considerably for the various food commodities and specific codes should be applied where appropriate, some general guidance is given in this section. Sections IV to X set down the general hygiene principles which apply throughout the food chain to the point of sale. Section IX also covers consumer information, recognizing the important role played by consumers in maintaining the safety and suitability of food.

There will inevitably be situations where some of the specific requirements contained in this document are not applicable. The fundamental question in every case is "what is necessary and appropriate on the grounds of the safety and suitability of food for consumption?"

The text indicates where such questions are likely to arise by using the phrases "where necessary" and "where appropriate". In practice, this means that, although the requirement is generally appropriate and reasonable, there will nevertheless be some situations where it is neither necessary nor appropriate on the grounds of food safety and suitability. In deciding whether a requirement is necessary or appropriate, an assessment of the risk should be made, preferably within the framework of the HACCP approach. This approach allows the requirements in this document to be flexibly and sensibly applied with a proper regard for the overall objectives of producing food which is safe and suitable for consumption. In so doing it takes into account the wide diversity of activities and varying degrees of risk involved in producing food. Additional guidance is available in specific food codes.

2.3   DEFINITIONS

For the purpose of this Code, the following expressions have the meaning stated:

Cleaning - the removal of soil, food residue, dirt, grease or other objectionable matter.

Contaminant - any biological or chemical agent, foreign matter, or other substances not intentionally added to food which may compromise food safety or suitability.

Contamination - the introduction or occurrence of a contaminant in food or food environment.

Disinfection - the reduction, by means of chemical agents and/or physical methods, of the number of micro-organisms in the environment, to a level that does not compromise food safety or suitability.

Establishment - any building or area in which food is handled and the surroundings under the control of the same management.

Food hygiene - all conditions and measures necessary to ensure the safety and suitability of food at all stages of the food chain.

Hazard - a biological, chemical or physical agent in, or condition of, food with the potential to cause an adverse health effect.

HACCP - a system which identifies, evaluates, and controls hazards which are significant for food safety.

Food handler - any person who directly handles packaged or unpackaged food, food equipment and utensils, or food contact surfaces and is therefore expected to comply with food hygiene requirements

Food safety - assurance that food will not cause harm to the consumer when it is prepared and/or eaten according to its intended use.

Food suitability - assurance that food is acceptable for human consumption according to its intended use.

Primary production - those steps in the food chain up to and including, for example, harvesting, slaughter, milking, fishing.

SECTION III - PRIMARY PRODUCTION

> ### Objectives:
>
> *Primary production should be managed in a way that ensures that food is safe and suitable for its intended use. Where necessary, this will include:*
>
> — *avoiding the use of areas where the environment poses a threat to the safety of food;*
>
> — *controlling contaminants, pests and diseases of animals and plants in such a way as not to pose a threat to food safety;*
>
> — *adopting practices and measures to ensure food is produced under appropriately hygienic conditions.*
>
> ### Rationale:
>
> *To reduce the likelihood of introducing a hazard which may adversely affect the safety of food, or its suitability for consumption, at later stages of the food chain.*

3.1   ENVIRONMENTAL HYGIENE

Potential sources of contamination from the environment should be considered. In particular, primary food production should not be carried on in areas where the presence of potentially harmful substances would lead to an unacceptable level of such substances in food.

3.2   HYGIENIC PRODUCTION OF FOOD SOURCES

The potential effects of primary production activities on the safety and suitability of food should be considered at all times. In particular, this includes identifying any specific points in such activities where a high probability of contamination may exist and taking specific measures to minimize that probability. The HACCP-based approach may assist in the taking of such measures - see *Hazard Analysis and Critical Control (HACCP) Point System and Guidelines for its Application* (Annex, page 31).

Producers should as far as practicable implement measures to:

- control contamination from air, soil, water, feedstuffs, fertilizers (including natural fertilizers), pesticides, veterinary drugs or any other agent used in primary production;

- control plant and animal health so that it does not pose a threat to human health through food consumption, or adversely affect the suitability of the product; and

- protect food sources from faecal and other contamination.

In particular, care should be taken to manage wastes, and store harmful substances appropriately. On-farm programmes which achieve specific food safety goals are becoming an important part of primary production and should be encouraged.

3.3   HANDLING, STORAGE AND TRANSPORT

Procedures should be in place to:

- sort food and food ingredients to segregate material which is evidently unfit for human consumption;

- dispose of any rejected material in a hygienic manner; and

- Protect food and food ingredients from contamination by pests, or by chemical, physical or microbiological contaminants or other objectionable substances during handling, storage and transport.

Care should be taken to prevent, so far as reasonably practicable, deterioration and spoilage through appropriate measures which may include controlling temperature, humidity, and/or other controls.

3.4   CLEANING, MAINTENANCE AND PERSONNEL HYGIENE AT PRIMARY PRODUCTION

Appropriate facilities and procedures should be in place to ensure that:

- any necessary cleaning and maintenance is carried out effectively; and
- an appropriate degree of personal hygiene is maintained.

SECTION IV - ESTABLISHMENT: DESIGN AND FACILITIES

Objectives:

Depending on the nature of the operations, and the risks associated with them, premises, equipment and facilities should be located, designed and constructed to ensure that:

- *contamination is minimized;*

- *design and layout permit appropriate maintenance, cleaning and disinfections and minimize air-borne contamination;*

- *surfaces and materials, in particular those in contact with food, are non-toxic in intended use and, where necessary, suitably durable, and easy to maintain and clean;*

- *where appropriate, suitable facilities are available for temperature, humidity and other controls; and*

- *there is effective protection against pest access and harbourage.*

Rationale:

Attention to good hygienic design and construction, appropriate location, and the provision of adequate facilities, is necessary to enable hazards to be effectively controlled.

4.1   LOCATION

4.1.1 ESTABLISHMENTS

Potential sources of contamination need to be considered when deciding where to locate food establishments, as well as the effectiveness of any reasonable measures that might be taken to protect food. Establishments should not be located anywhere where, after considering such protective measures, it is clear that there will remain a threat to food safety or suitability. In particular, establishments should normally be located away from:

-    environmentally polluted areas and industrial activities which pose a serious threat of contaminating food;
-    areas subject to flooding unless sufficient safeguards are provided;
-    areas prone to infestations of pests;
-    areas where wastes, either solid or liquid, cannot be removed effectively.

4.1.2 EQUIPMENT

Equipment should be located so that it:

-    permits adequate maintenance and cleaning;
-    functions in accordance with its intended use; and
-    facilitates good hygiene practices, including monitoring.

4.2 PREMISES AND ROOMS

4:2.1 DESIGN AND LAYOUT

Where appropriate, the internal design and layout of food establishments should permit good food hygiene practices, including protection against cross-contamination between and during operations by foodstuffs.

4.2.2 INTERNAL STRUCTURES AND FITTINGS

Structures within food establishments should be soundly built of durable materials and be easy to maintain, clean and where appropriate, able to be disinfected. In particular the following specific conditions should be satisfied where necessary to protect the safety and suitability of food:

-    the surfaces of walls, partitions and floors should be made of impervious materials with no toxic effect in intended use;
-    walls and partitions should have a smooth surface up to a height appropriate to the operation;
-    floors should be constructed to allow adequate drainage and cleaning;
-    ceilings and overhead fixtures should be constructed and finished to minimize the build up of dirt and condensation, and the shedding of particles;

- windows should be easy to clean, be constructed to minimize the build up of dirt and where necessary, be fitted with removable and cleanable insect-proof screens. Where necessary, windows should be fixed;

- doors should have smooth, non-absorbent surfaces, and be easy to clean and, where necessary, disinfect;

- working surfaces that come into direct contact with food should be in sound condition, durable and easy to clean, maintain and disinfect. They should be made of smooth, non-absorbent materials, and inert to the food, to detergents and disinfectants under normal operating conditions.

4.2.3 TEMPORARY/MOBILE PREMISES AND VENDING MACHINES

Premises and structures covered here include market stalls, mobile sales and street vending vehicles, temporary premises in which food is handled such as tents and marquees.

Such premises and structures should be sited, designed and constructed to avoid, as far as reasonably practicable, contaminating food and harbouring pests.

In applying these specific conditions and requirements, any food hygiene hazards associated with such facilities should be adequately controlled to ensure the safety and suitability of food.

4.3 EQUIPMENT

4.3.1 GENERAL

Equipment and containers (other than once-only use containers and packaging) coming into contact with food, should be designed and constructed to ensure that, where necessary, they can be adequately cleaned, disinfected and maintained to avoid the contamination of food. Equipment and containers should be made of materials with no toxic effect in intended use. Where necessary, equipment should be durable and movable or capable of being disassembled to allow for maintenance, cleaning, disinfection, monitoring and, for example, to facilitate inspection for pests.

4.3.2 FOOD CONTROL AND MONITORING EQUIPMENT

In addition to the general requirements in paragraph 4.3.1, equipment used to cook, heat treat, cool, store or freeze food should be designed to achieve the required food temperatures as rapidly as necessary in the interests of food safety and suitability, and maintain them effectively. Such equipment should also be designed to allow temperatures to be monitored and controlled. Where necessary, such equipment should have effective means of controlling and monitoring humidity, air-flow and any other characteristic likely to have a detrimental effect on the safety or suitability of food. These requirements are intended to ensure that:

- harmful or undesirable micro-organisms or their toxins are eliminated or reduced to safe levels or their survival and growth are effectively controlled;

- where appropriate, critical limits established in HACCP-based plans can be monitored; and

- temperatures and other conditions necessary to food safety and suitability can be rapidly achieved and maintained.

4.3.3 CONTAINERS FOR WASTE AND INEDIBLE SUBSTANCES

Containers for waste, by-products and inedible or dangerous substances, should be specifically identifiable, suitably constructed and, where appropriate, made of impervious material. Containers used to hold dangerous substances should be identified and, where appropriate, be lockable to prevent malicious or accidental contamination of food.

4.4 FACILITIES

4.4.1 WATER SUPPLY

An adequate supply of potable water with appropriate facilities for its storage, distribution and temperature control, should be available whenever necessary to ensure the safety and suitability of food.

Potable water should be as specified in the latest edition of WHO Guidelines for Drinking Water Quality, or water of a higher standard. Non-potable water (for use in, for example, fire control, steam production, refrigeration and other similar purposes where it would not contaminate food), shall have a separate system. Non-potable water systems shall be identified and shall not connect with, or allow reflux into, potable water systems.

4.4.2 DRAINAGE AND WASTE DISPOSAL

Adequate drainage and waste disposal systems and facilities should be provided. They should be designed and constructed so that the risk of contaminating food or the potable water supply is avoided.

4.4.3 CLEANING

Adequate facilities, suitably designated, should be provided for cleaning food, utensils and equipment. Such facilities should have an adequate supply of hot and cold potable water where appropriate.

4.4.4 PERSONNEL HYGIENE FACILITIES AND TOILETS

Personnel hygiene facilities should be available to ensure that an appropriate degree of personal hygiene can be maintained and to avoid contaminating food. Where appropriate, facilities should include:

- adequate means of hygienically washing and drying hands, including wash basins and a supply of hot and cold (or suitably temperature controlled) water;
- lavatories of appropriate hygienic design; and
- adequate changing facilities for personnel.

Such facilities should be suitably located and designated.

4.4.5 TEMPERATURE CONTROL

Depending on the nature of the food operations undertaken, adequate facilities should be available for heating, cooling, cooking, refrigerating and freezing food, for storing refrigerated or frozen foods, monitoring food temperatures, and when necessary, controlling ambient temperatures to ensure the safety and suitability of food.

4.4.6 AIR QUALITY AND VENTILATION

Adequate means of natural or mechanical ventilation should be provided, in particular to:

- minimize air-borne contamination of food, for example, from aerosols and condensation droplets;
- control ambient temperatures;
- control odours which might affect the suitability of food; and

       — control humidity, where necessary, to ensure the safety and suitability of food.

Ventilation systems should be designed and constructed so that air does not flow from contaminated areas to clean areas and, where necessary, they can be adequately maintained and cleaned.

4.4.7 LIGHTING

Adequate natural or artificial lighting should be provided to enable the undertaking to operate in a hygienic manner. Where necessary, lighting should not be such that the resulting colour is misleading. The intensity should be adequate to the nature of the operation. Lighting fixtures should, where appropriate, be protected to ensure that food is not contaminated by breakages.

4.4.8 STORAGE

Where necessary, adequate facilities for the storage of food, ingredients and non-food chemicals (e.g. cleaning materials, lubricants, fuels) should be provided.

Where appropriate, food storage facilities should be designed and constructed to:

       — permit adequate maintenance and cleaning;

       — avoid pest access and harbourage;

       — enable food to be effectively protected from contamination during storage; and

       — where necessary, provide an environment which minimizes the deterioration of food (e.g. by temperature and humidity control).

The type of storage facilities required will depend on the nature of the food. Where necessary, separate, secure storage facilities for cleaning materials and hazardous substances should be provided.

SECTION V - CONTROL OF OPERATION

> **Objective:**
>
> *To produce food which is safe and suitable for human consumption by:*
>
> – *formulating design requirements with respect to raw materials, composition, processing, distribution, and consumer use to be met in the manufacture and handling of specific food items; and*
>
> – *designing, implementing, monitoring and reviewing effective control systems.*
>
> **Rationale:**
>
> *To reduce the risk of unsafe food by taking preventive measures to assure the safety and suitability of food at an appropriate stage in the operation by controlling food hazards.*

5.1 CONTROL OF FOOD HAZARDS

Food business operators should control food hazards through the use of systems such as HACCP. They should:

- **identify** any steps in their operations which are critical to the safety of food;

- **implement** effective control procedures at those steps;

- **monitor** control procedures to ensure their continuing effectiveness; and

- **review** control procedures periodically, and whenever the operations change.

These systems should be applied throughout the food chain to control food hygiene throughout the shelf-life of the product through proper product and process design.

Control procedures may be simple, such as checking stock rotation calibrating equipment, or correctly loading refrigerated display units. In some cases a system based on expert advice, and involving documentation, may be appropriate. A model of such a food safety system is described in *Hazard*

Analysis and Critical Control (HACCP) System and Guidelines for its Application (Annex).

5.2   KEY ASPECTS OF HYGIENE CONTROL SYSTEMS

5.2.1   TIME AND TEMPERATURE CONTROL

Inadequate food temperature control is one of the most common causes of foodborne illness or food spoilage. Such controls include time and temperature of cooking, cooling, processing and storage. Systems should be in place to ensure that temperature is controlled effectively where it is critical to the safety and suitability of food.

Temperature control systems should take into account:

– the nature of the food, e.g. its water activity, pH, and likely initial level and types of micro-organisms;

– the intended shelf-life of the product;

– the method of packaging and processing; and

– how the product is intended to be used, e.g. further cooking/processing or ready-to-eat.

Such systems should also specify tolerable limits for time and temperature variations.

Temperature recording devices should be checked at regular intervals and tested for accuracy.

5.2.2   SPECIFIC PROCESS STEPS

Other steps which contribute to food hygiene may include, for example:

– chilling

– thermal processing

– irradiation

– drying

– chemical preservation

– vacuum or modified atmospheric packaging

5.2.3 MICROBIOLOGICAL AND OTHER SPECIFICATIONS

Management systems described in paragraph 5.1 offer an effective way of ensuring the safety and suitability of food. Where microbiological, chemical or physical specifications are used in any food control system, such specifications should be based on sound scientific principles and state, where appropriate, monitoring procedures, analytical methods and action limits.

5.2.4 MICROBIOLOGICAL CROSS-CONTAMINATION

Pathogens can be transferred from one food to another, either by direct contact or by food handlers, contact surfaces or the air. Raw, unprocessed food should be effectively separated, either physically or by time, from ready-to-eat foods, with effective intermediate cleaning and where appropriate disinfection.

Access to processing areas may need to be restricted or controlled. Where risks are particularly high, access to processing areas should be only via a changing facility. Personnel may need to be required to put on clean protective clothing including footwear and wash their hands before entering.

Surfaces, utensils, equipment, fixtures and fittings should be thoroughly cleaned and where necessary disinfected after raw food, particularly meat and poultry, has been handled or processed.

5.2.5 PHYSICAL AND CHEMICAL CONTAMINATION

Systems should be in place to prevent contamination of foods by foreign bodies such as glass or metal shards from machinery, dust, harmful fumes and unwanted chemicals. In manufacturing and processing, suitable detection or screening devices should be used where necessary.

5.3 INCOMING MATERIAL REQUIREMENTS

No raw material or ingredient should be accepted by an establishment if it is known to contain parasites, undesirable micro-organisms, pesticides, veterinary drugs or toxic, decomposed or extraneous substances which would not be reduced to an acceptable level by normal sorting and/or processing. Where appropriate, specifications for raw materials should be identified and applied.

Raw materials or ingredients should, where appropriate, be inspected and sorted before processing. Where necessary, laboratory tests should be made to establish fitness for use. Only sound, suitable raw materials or ingredients should be used.

Stocks of raw materials and ingredients should be subject to effective stock rotation.

5.4   PACKAGING

Packaging design and materials should provide adequate protection for products to minimize contamination, prevent damage, and accommodate proper labelling. Packaging materials or gases where used must be non-toxic and not pose a threat to the safety and suitability of food under the specified conditions of storage and use.   Where appropriate, reusable packaging should be suitably durable, easy to clean and, where necessary, disinfect.

5.5   WATER

5.5.1   IN CONTACT WITH FOOD

Only potable water, should be used in food handling and processing, with the following exceptions:

- – for steam production, fire control and other similar purposes not connected with food; and
- – in certain food processes, e.g. chilling, and in food handling areas, provided this does not constitute a hazard to the safety and suitability of food (e.g. the use of clean sea water).

Water recirculated for reuse should be treated and maintained in such a condition that no risk to the safety and suitability of food results from its use. The treatment process should be effectively monitored.   Recirculated water which has received no further treatment and water recovered from processing of food by evaporation or drying may be used, provided its use does not constitute a risk to the safety and suitability of food.

5.5.2   AS AN INGREDIENT

Potable water should be used wherever necessary to avoid food contamination.

5.5.3   ICE AND STEAM

Ice should be made from water that complies with section 4.4.1.   Ice and steam should be produced, handled and stored to protect them from contamination.

Steam used in direct contact with food or food contact surfaces should not constitute a threat to the safety and suitability of food.

5.6   MANAGEMENT AND SUPERVISION

The type of control and supervision needed will depend on the size of the business, the nature of its activities and the types of food involved. Managers and supervisors should have enough knowledge of food hygiene principles and practices to be able to judge potential risks, take appropriate preventive and corrective action, and ensure that effective monitoring and supervision takes place.

5.7   DOCUMENTATION AND RECORDS

Where necessary, appropriate records of processing, production and distribution should be kept and retained for a period that exceeds the shelf-life of the product. Documentation can enhance the credibility and effectiveness of the food safety control system.

5.8   RECALL PROCEDURES

Managers should ensure effective procedures are in place to deal with any food safety hazard and to enable the complete, rapid recall of any implicated lot of the finished food from the market. Where a product has been withdrawn because of an immediate health hazard, other products which are produced under similar conditions, and which may present a similar hazard to public health, should be evaluated for safety and may need to be withdrawn. The need for public warnings should be considered.

Recalled products should be held under supervision until they are destroyed, used for purposes other than human consumption, determined to be safe for human consumption, or reprocessed in a manner to ensure their safety.

SECTION VI - ESTABLISHMENT: MAINTENANCE AND SANITATION

> *Objective:*
>
> *To establish effective systems to:*
>
> – *ensure adequate and appropriate maintenance and cleaning;*
>
> – *control pests;*
>
> – *manage waste; and*
>
> – *monitor effectiveness of maintenance and sanitation procedures.*
>
> *Rationale:*
>
> *To facilitate the continuing effective control of food hazards, pests, and other agents likely to contaminate food.*

6.1 MAINTENANCE AND CLEANING

6.1.1 GENERAL

Establishments and equipment should be kept in an appropriate state of repair and condition to:

- facilitate all sanitation procedures;
- function as intended, particularly at critical steps (see paragraph 5.1);
- prevent contamination of food, e.g. from metal shards, flaking plaster, debris and chemicals.

Cleaning should remove food residues and dirt which may be a source of contamination. The necessary cleaning methods and materials will depend on the nature of the food business. Disinfection may be necessary after cleaning.

Cleaning chemicals should be handled and used carefully and in accordance with manufacturers' instructions and stored, where necessary, separated from food, in clearly identified containers to avoid the risk of contaminating food.

6.1.2 CLEANING PROCEDURES AND METHODS

Cleaning can be carried out by the separate or the combined use of physical methods, such as heat, scrubbing, turbulent flow, vacuum cleaning or other methods that avoid the use of water, and chemical methods using detergents, alkalis or acids.

Cleaning procedures will involve, where appropriate:

-   removing gross debris from surfaces;
-   applying a detergent solution to loosen soil and bacterial film and hold them in solution or suspension;
-   rinsing with water which complies with section 4, to remove loosened soil and residues of detergent;
-   dry cleaning or other appropriate methods for removing and collecting residues and debris; and   •
-   where necessary, disinfection with subsequent rinsing unless the manufacturers' instructions indicate on a scientific basis that rinsing is not required.

6.2 CLEANING PROGRAMMES

Cleaning and disinfection programmes should ensure that all parts of the establishment are appropriately clean, and should include the cleaning of cleaning equipment.

Cleaning and disinfection programmes should be continually and effectively monitored for their suitability and effectiveness and where necessary, documented.

Where written cleaning programmes are used, they should specify:

-   areas, items of equipment and utensils to be cleaned;
-   responsibility for particular tasks;
-   method and frequency of cleaning; and
-   monitoring arrangements.

Where appropriate, programmes should be drawn up in consultation with relevant specialist expert advisors.

6.3   PEST CONTROL SYSTEMS

6.3.1 GENERAL

Pests pose a major threat to the safety and suitability of food. Pest infestations can occur where there are breeding sites and a supply of food. Good hygiene practices should be employed to avoid creating an environment conducive to pests. Good sanitation, inspection of incoming materials and good monitoring can minimize the likelihood of infestation and thereby limit the need for pesticides. [Insert reference to FAO document dealing with Integrated Pest Management].

6.3.2 PREVENTING ACCESS

Buildings should be kept in good repair and condition to prevent pest access and to eliminate potential breeding sites. Holes, drains and other places where pests are likely to gain access should be kept sealed. Wire mesh screens, for example on open windows, doors and ventilators, will reduce the problem of pest entry. Animals should, wherever possible, be excluded from the grounds of factories and food processing plants.

6.3.3 HARBOURAGE AND INFESTATION

The availability of food and water encourages pest harbourage and infestation. Potential food sources should be stored in pest-proof containers and/or stacked above the ground and away from walls. Areas both inside and outside food premises should be kept clean. Where appropriate, refuse should be stored in covered, pest-proof containers.

6.3.4 MONITORING AND DETECTION

Establishments and surrounding areas should be regularly examined for evidence of infestation.

6.3.5 ERADICATION

Pest infestations should be dealt with immediately and without adversely affecting food safety or suitability. Treatment with chemical, physical or biological agents should be carried out without posing a threat to the safety or suitability of food.

6.4   WASTE MANAGEMENT

Suitable provision must be made for the removal and storage of waste. Waste must not be allowed to accumulate in food handling, food storage, and other working areas and the adjoining environment except so far as is unavoidable for the proper functioning of the business.

Waste stores must be kept appropriately clean.

6.5   MONITORING EFFECTIVENESS

Sanitation systems should be monitored for effectiveness, periodically verified by means such as audit pre-operational inspections or, where appropriate, microbiological sampling of environment and food contact surfaces and regularly reviewed and adapted to reflect changed circumstances.

SECTION VII - ESTABLISHMENT: PERSONAL HYGIENE

Objectives:

To ensure that those who come directly or indirectly into contact with food are not likely to contaminate food by:

- *maintaining an appropriate degree of personal cleanliness;*

- *behaving and operating in an appropriate manner.*

Rationale:

People who do not maintain an appropriate degree of personal cleanliness, who have certain illnesses or conditions or who behave inappropriately, can contaminate food and transmit illness to consumers.

7.1 HEALTH STATUS

People known, or suspected, to be suffering from, or to be a carrier of a disease or illness likely to be transmitted through food, should not be allowed to enter any food handling area if there is a likelihood of their contaminating food. Any person so affected should immediately report illness or symptoms of illness to the management.

Medical examination of a food handler should be carried out if clinically or epidemiologically indicated.

7.2 ILLNESS AND INJURIES

Conditions which should be reported to management so that any need for medical examination and/or possible exclusion from food handling can be considered, include:

- jaundice
- diarrhoea
- vomiting
- fever
- sore throat with fever

- visibly infected skin lesions (boils, cuts, etc.)
- discharges from the ear, eye or nose

7.3 PERSONAL CLEANLINESS

Food handlers should maintain a high degree of personal cleanliness and, where appropriate, wear suitable protective clothing, head covering, and footwear. Cuts and wounds, where personnel are permitted to continue working, should be covered by suitable waterproof dressings.

Personnel should always wash their hands when personal cleanliness may affect food safety, for example:

- at the start of food handling activities;
- immediately after using the toilet; and
- after handling raw food or any contaminated material, where this could result in contamination of other food items; they should avoid handling ready-to-eat food, where appropriate.

7.4 PERSONAL BEHAVIOUR

People engaged in food handling activities should refrain from behaviour which could result in contamination of food, for example:

- smoking;
- spitting;
- chewing or eating;
- sneezing or coughing over unprotected food.

Personal effects such as jewellery, watches, pins or other items should not be worn or brought into food handling areas if they pose a threat to the safety and suitability of food.

7.5 VISITORS

Visitors to food manufacturing, processing or handling areas should, where appropriate, wear protective clothing and adhere to the other personal hygiene provisions in this section.

SECTION VIII - TRANSPORTATION

Objectives:

Measures should be taken where necessary to:

- *protect food from potential sources of contamination;*

- *protect food from damage likely to render the food unsuitable for consumption; and*

- *provide an environment which effectively controls the growth of pathogenic or spoilage micro-organisms and the production of toxins in food.*

Rationale:

Food may become contaminated, or may not reach its destination in a suitable condition for consumption, unless effective control measures are taken during transport, even where adequate hygiene control measures have been taken earlier in the food chain.

8.1 GENERAL

Food must be adequately protected during transport. The type of conveyances or containers required depends on the nature of the food and the conditions under which it has to be transported.

8.2 REQUIREMENTS

Where necessary, conveyances and bulk containers should be designed and constructed so that they:

- do not contaminate foods or packaging;

- can be effectively cleaned and, where necessary, disinfected;

- permit effective separation of different foods or foods from non-food items where necessary during transport;

- provide effective protection from contamination, including dust and fumes;

- can effectively maintain the temperature, humidity, atmosphere and other conditions necessary to protect food from harmful or undesirable microbial growth and deterioration likely to render it unsuitable for consumption; and

– allow any necessary temperature, humidity and other conditions to be checked.

8.3   USE AND MAINTENANCE

Conveyances and containers for transporting food should be kept in an appropriate state of cleanliness, repair and condition. Where the same conveyance or container is used for transporting different foods, or non-foods, effective cleaning and, where necessary, disinfection should take place between loads.

Where appropriate, particularly in bulk transport, containers and conveyances should be designated and marked for food use only and be used only for that purpose.

SECTION IX - PRODUCT INFORMATION AND CONSUMER AWARENESS

Objectives:

Products should bear appropriate information to ensure that:

- *adequate and accessible information is available to the next person in the food chain to enable them to handle, store, process, prepare and display the product safely and correctly;*

- *the lot or batch can be easily identified and recalled if necessary.*

Consumers should have enough knowledge of food hygiene to enable them to:

- *understand the importance of product information;*

- *make informed choices appropriate to the individual; and*

- *prevent contamination and growth or survival of foodborne pathogens by storing, preparing and using it correctly.*

Information for industry or trade users should be clearly distinguishable from consumer information, particularly on food labels.

Rationale:

Insufficient product information, and/or inadequate knowledge of general food hygiene, can lead to products being mishandled at later stages in the food chain. Such mishandling can result in illness, or products becoming unsuitable for consumption, even where adequate hygiene control measures have been taken earlier in the food chain.

9.1 LOT IDENTIFICATION

Lot identification is essential in product recall and also helps effective stock rotation. Each container of food should be permanently marked to identify the producer and the lot. Codex General Standard for the Labelling of Prepackaged Foods (CODEX STAN 1-1985) applies.

9.2   PRODUCT INFORMATION

All food products should be accompanied by or bear adequate information to enable the next person in the food chain to handle, display, store and prepare and use the product safely and correctly.

9.3   LABELLING

Prepackaged foods should be labelled with clear instructions to enable the next person in the food chain to handle, display, store and use the product safely. Codex General Standard for the Labelling of Prepackaged Foods (CODEX STAN 1-1985) applies.

9.4   CONSUMER EDUCATION

Health education programmes should cover general food hygiene. Such programmes should enable consumers to understand the importance of any product information and to follow any instructions accompanying products, and make informed choices. In particular consumers should be informed of the relationship between time/temperature control and foodborne illness.

SECTION X - TRAINING

Objective:

Those engaged in food operations who come directly or indirectly into contact with food should be trained, and/or instructed in food hygiene to a level appropriate to the operations they are to perform.

Rationale:

Training is fundamentally important to any food hygiene system.

Inadequate hygiene training, and/or instruction and supervision of all people involved in food related activities pose a potential threat to the safety of food and its suitability for consumption.

10.1   AWARENESS AND RESPONSIBILITIES

Food hygiene training is fundamentally important. All personnel should be aware of their role and responsibility in protecting food from contamination or deterioration. Food handlers should have the necessary knowledge and skills to enable them to handle food hygienically. Those who handle strong cleaning chemicals or other potentially hazardous chemicals should be instructed in safe handling techniques.

10.2 TRAINING PROGRAMMES

Factors to take into account in assessing the level of training required include:

- the nature of the food, in particular its ability to sustain growth of pathogenic or spoilage micro-organisms;
- the manner in which the food is handled and packed, including the probability of contamination;
- the extent and nature of processing or further preparation before final consumption;
- the conditions under which the food will be stored; and
- the expected length of time before consumption.

10.3 INSTRUCTION AND SUPERVISION

Periodic assessments of the effectiveness of training and instruction programmes should be made, as well as routine supervision and checks to ensure that procedures are being carried out effectively.

Managers and supervisors of food processes should have the necessary knowledge of food hygiene principles and practices to be able to judge potential risks and take the necessary action to remedy deficiencies.

10.4 REFRESHER TRAINING

Training programmes should be routinely reviewed and updated where necessary. Systems should be in place to ensure that food handlers remain aware of all procedures necessary to maintain the safety and suitability of food.

HAZARD ANALYSIS AND CRITICAL CONTROL POINT (HACCP) SYSTEM AND GUIDELINES FOR ITS APPLICATION

ANNEX

PREAMBLE

The first section of this document sets out the principles of the Hazard Analysis and Critical Control Point (HACCP) system adopted by the Codex Alimentarius Commission. The second section provides general guidance for the application of the system while recognizing that the details of application may vary depending on the circumstances of the food operation.[1]

The HACCP system, which is science based and systematic, identifies specific hazards and measures for their control to ensure the safety of food. HACCP is a tool to assess hazards and establish control systems that focus on prevention rather than relying mainly on end-product testing. Any HACCP system is capable of accommodating change, such as advances in equipment design, processing procedures or technological developments.

HACCP can be applied throughout the food chain from primary production to final consumption and its implementation should be guided by scientific evidence of risks to human health. As well as enhancing food safety, implementation of HACCP can provide other significant benefits. In addition, the application of HACCP systems can aid inspection by regulatory authorities and promote international trade by increasing confidence in food safety.

The successful application of HACCP requires the full commitment and involvement of management and the work force. It also requires a multidisciplinary approach; this multidisciplinary approach should include, when appropriate, expertise in agronomy, veterinary health, production, microbiology, medicine, public health, food technology, environmental health, chemistry and engineering, according to the particular study. The application of HACCP is compatible with the implementation of quality management systems,

---

[1] The Principles of the HACCP System set the basis for the requirements for the application of HACCP, while the Guidelines for the Application provide general guidance for practical application.

such as the ISO 9000 series, and is the system of choice in the management of food safety within such systems.

While the application of HACCP to food safety was considered here, the concept can be applied to other aspects of food quality.

DEFINITIONS

Control (verb): To take all necessary actions to ensure and maintain compliance with criteria established in the HACCP plan.

Control (noun): The state wherein correct procedures are being followed and criteria are being met.

Control measure: Any action and activity that can be used to prevent or eliminate a food safety hazard or reduce it to an acceptable level.

Corrective action: Any action to be taken when the results of monitoring at the CCP indicate a loss of control.

Critical Control Point (CCP): A step at which control can be applied and is essential to prevent or eliminate a food safety hazard or reduce it to an acceptable level.

Critical limit: A criterion which separates acceptability from unacceptability.

Deviation: Failure to meet a critical limit.

Flow diagram: A systematic representation of the sequence of steps or operations used in the production or manufacture of a particular food item.

HACCP: A system which identifies, evaluates, and controls hazards which are significant for food safety.

HACCP plan: A document prepared in accordance with the principles of HACCP to ensure control of hazards which are significant for food safety in the segment of the food chain under consideration.

Hazard: A biological, chemical or physical agent in, or condition of, food with the potential to cause an adverse health effect.

Hazard analysis: The process of collecting and evaluating information on hazards and conditions leading to their presence to decide which are significant for food safety and therefore should be addressed in the HACCP plan.

Monitor: The act of conducting a planned sequence of observations or measurements of control parameters to assess whether a CCP is under control.

Step: A point, procedure, operation or stage in the food chain including raw materials, from primary production to final consumption.

Validation: Obtaining evidence that the elements of the HACCP plan are effective.

Verification: The application of methods, procedures, tests and other evaluations, in addition to monitoring to determine compliance with the HACCP plan.

PRINCIPLES OF THE HACCP SYSTEM

The HACCP system consists of the following seven principles:

PRINCIPLE 1

Conduct a hazard analysis.

PRINCIPLE 2

Determine the Critical Control Points (CCPs).

PRINCIPLE 3

Establish critical limit(s).

PRINCIPLE 4

Establish a system to monitor control of the CCP.

PRINCIPLE 5

Establish the corrective action to be taken when monitoring indicates that a particular CCP is not under control.

PRINCIPLE 6

Establish procedures for verification to confirm that the HACCP system is working effectively.

PRINCIPLE 7

Establish documentation concerning all procedures and records appropriate to these principles and their application.

GUIDELINES FOR THE APPLICATION OF THE HACCP SYSTEM

Prior to application of HACCP to any sector of the food chain, that sector should have in place prerequisite programs such as good hygienic practices according to the Codex General Principles of Food Hygiene, the appropriate Codex Codes of Practice, and appropriate food safety requirements. These prerequisite programs to HACCP, including training, should be well established, fully operational and verified in order to facilitate the successful application and implementation of the HACCP system.

For all types of food business, management awareness and commitment is necessary for implementation of an effective HACCP system. The effectiveness will also rely upon management and employees having the appropriate HACCP knowledge and skills.

During hazard identification, evaluation, and subsequent operations in designing and applying HACCP systems, consideration must be given to the impact of raw materials, ingredients, food manufacturing practices, role of manufacturing processes to control hazards, likely end-use of the product, categories of consumers of concern, and epidemiological evidence relative to food safety.

The intent of the HACCP system is to focus control at Critical Control Points (CCPs). Redesign of the operation should be considered if a hazard which must be controlled is identified but no CCPs are found.

HACCP should be applied to each specific operation separately. CCPs identified in any given example in any Codex Code of Hygienic Practice might not be the only ones identified for a specific application or might be of a different nature. The HACCP application should be reviewed and necessary changes made when any modification is made in the product, process, or any step.

The application of the HACCP principles should be the responsibility of each individual businesses. However, it is recognised by governments and businesses that there may be obstacles that hinder the effective application of the HACCP principles by individual business. This is particularly relevant in small and/or less developed businesses. While it is recognized that when applying HACCP, flexibility appropriate to the business is important, all seven principles must be applied in the HACCP system. This flexibility should take into account the nature and size of the operation, including the human and financial resources, infrastructure, processes, knowledge and practical constraints.

Small and/or less developed businesses do not always have the resources and the necessary expertise on site for the development and implementation of an effective HACCP plan. In such situations, expert advice should be obtained from other sources, which may include: trade and industry associations, independent experts and regulatory authorities. HACCP literature and especially sector-specific HACCP guides can be valuable. HACCP guidance developed by experts relevant to the process or type of operation may provide a useful tool for businesses in designing and implementing the HACCP plan. Where businesses are using expertly developed HACCP guidance, it is essential that it is specific to the foods and/or processes under consideration. More detailed information on the obstacles in implementing HACCP, particularly in reference to SLDBs, and recommendations in resolving these obstacles, can be found in "Obstacles to the Application of HACCP, Particularly in Small and Less Developed Businesses, and Approaches to Overcome Them" (document in preparation by FAO/WHO).

The efficacy of any HACCP system will nevertheless rely on management and employees having the appropriate HACCP knowledge and skills, therefore ongoing training is necessary for all levels of employees and managers, as appropriate.

APPLICATION

The application of HACCP principles consists of the following tasks as identified in the Logic Sequence for Application of HACCP (Diagram 1).

1.   Assemble HACCP team

The food operation should assure that the appropriate product specific knowledge and expertise is available for the development of an effective HACCP plan.   Optimally, this may be accomplished by assembling a multidisciplinary team.   Where such expertise is not available on site, expert advice should be obtained from other sources, such as, trade and industry associations, independent experts, regulatory authorities, HACCP literature and HACCP guidance (including sector-specific HACCP guides).   It may be possible that a well-trained individual with access to such guidance is able to implement HACCP in-house.   The scope of the HACCP plan should be identified.   The scope should describe which segment of the food chain is involved and the general classes of hazards to be addressed (e.g. does it cover all classes of hazards or only selected classes).

2.   Describe product

A full description of the product should be drawn up, including relevant safety information such as: composition, physical/chemical structure (including A_w, pH, etc.), microcidal/static treatments (heat-treatment, freezing, brining, smoking, etc.), packaging, durability and storage conditions and method of distribution.   Within businesses with multiple products, for example, catering operations, it may be effective to group products with similar characteristics or processing steps, for the purpose of development of the HACCP plan.

3.   Identify intended use

The intended use should be based on the expected uses of the product by the end user or consumer.   In specific cases, vulnerable groups of the population, e.g. institutional feeding, may have to be considered.

4.   Construct flow diagram

The flow diagram should be constructed by the HACCP team (see also paragraph 1 above).   The flow diagram should cover all steps in the operation for a specific product.   The same flow diagram may be used for a number of products that are manufactured using similar processing steps.   When applying HACCP to a given operation, consideration should be given to steps preceding and following the specified operation.

5.   On-site confirmation of flow diagram

Steps must be taken to confirm the processing operation against the flow diagram during all stages and hours of operation and amend the flow diagram where appropriate. The confirmation of the flow diagram should be performed by a person or persons with sufficient knowledge of the processing operation.

6.   List all potential hazards associated with each step, conduct a hazard analysis, and consider any measures to control identified hazards

(SEE PRINCIPLE 1)

The HACCP team (see "assemble HACCP team" above) should list all of the hazards that may be reasonably expected to occur at each step according to the scope from primary production, processing, manufacture, and distribution until the point of consumption.

The HACCP team (see "assemble HACCP team") should next conduct a hazard analysis to identify for the HACCP plan which hazards are of such a nature that their elimination or reduction to acceptable levels is essential to the production of a safe food.

In conducting the hazard analysis, wherever possible the following should be included:

– the likely occurrence of hazards and severity of their adverse health effects;

– the qualitative and/or quantitative evaluation of the presence of hazards;

– survival or multiplication of microorganisms of concern;

– production or persistence in foods of toxins, chemicals or physical agents; and,

– conditions leading to the above.

Consideration should be given to what control measures, if any exist, can be applied for each hazard.

More than one control measure may be required to control a specific hazard(s) and more than one hazard may be controlled by a specified control measure.

7. Determine Critical Control Points

(SEE PRINCIPLE 2)[2]

There may be more than one CCP at which control is applied to address the same hazard. The determination of a CCP in the HACCP system can be facilitated by the application of a decision tree (e.g. Diagram 2), which indicates a logic reasoning approach. Application of a decision tree should be flexible, given whether the operation is for production, slaughter, processing, storage, distribution or other. It should be used for guidance when determining CCPs. This example of a decision tree may not be applicable to all situations. Other approaches may be used. Training in the application of the decision tree is recommended.

If a hazard has been identified at a step where control is necessary for safety, and no control measure exists at that step, or any other, then the product or process should be modified at that step, or at any earlier or later stage, to include a control measure.

8. Establish critical limits for each CCP

(SEE PRINCIPLE 3)

Critical limits must be specified and validated for each Critical Control Point. In some cases more than one critical limit will be elaborated at a particular step. Criteria often used include measurements of temperature, time, moisture level, pH, A_w, available chlorine, and sensory parameters such as visual appearance and texture.

Where HACCP guidance developed by experts has been used to establish the critical limits, care should be taken to ensure that these limits fully apply to the specific operation, product or groups of products under consideration. These critical limits should be measurable.

---

[2]   Since the publication of the decision tree by Codex, its use has been implemented many times for training purposes. In many instances, while this tree has been useful to explain the logic and depth of understanding needed to determine CCPs, it is not specific to all food operations, e.g. slaughter, and therefore it should be used in conjunction with professional judgement, and modified in some cases.

9.   Establish a monitoring system for each CCP

(SEE PRINCIPLE 4)

Monitoring is the scheduled measurement or observation of a CCP relative to its critical limits. The monitoring procedures must be able to detect loss of control at the CCP. Further, monitoring should ideally provide this information in time to make adjustments to ensure control of the process to prevent violating the critical limits. Where possible, process adjustments should be made when monitoring results indicate a trend towards loss of control at a CCP. The adjustments should be taken before a deviation occurs. Data derived from monitoring must be evaluated by a designated person with knowledge and authority to carry out corrective actions when indicated. If monitoring is not continuous, then the amount or frequency of monitoring must be sufficient to guarantee the CCP is in control. Most monitoring procedures for CCPs will need to be done rapidly because they relate to on-line processes and there will not be time for lengthy analytical testing. Physical and chemical measurements are often preferred to microbiological testing because they may be done rapidly and can often indicate the microbiological control of the product.

All records and documents associated with monitoring CCPs must be signed by the person(s) doing the monitoring and by a responsible reviewing official(s) of the company.

10.   Establish corrective actions

(SEE PRINCIPLE 5)

Specific corrective actions must be developed for each CCP in the HACCP system in order to deal with deviations when they occur.

The actions must ensure that the CCP has been brought under control. Actions taken must also include proper disposition of the affected product. Deviation and product disposition procedures must be documented in the HACCP record keeping.

11.   Establish verification procedures

(SEE PRINCIPLE 6)

Establish procedures for verification. Verification and auditing methods, procedures and tests, including random sampling and analysis, can be used to determine if the HACCP system is working correctly. The frequency of verification should be sufficient to confirm that the HACCP system is working effectively.

Verification should be carried out by someone other than the person who is responsible for performing the monitoring and corrective actions. Where certain verification activities cannot be performed in house, verification should be performed on behalf of the business by external experts or qualified third parties.

Examples of verification activities include:

- Review of the HACCP system and plan and its records;
- Review of deviations and product dispositions;
- Confirmation that CCPs are kept under control.

Where possible, validation activities should include actions to confirm the efficacy of all elements of the HACCP system.

12. *Establish Documentation and Record Keeping*

(SEE PRINCIPLE 7)

Efficient and accurate record keeping is essential to the application of a HACCP system. HACCP procedures should be documented. Documentation and record keeping should be appropriate to the nature and size of the operation and sufficient to assist the business to verify that the HACCP controls are in place and being maintained. Expertly developed HACCP guidance materials (e.g. sector-specific HACCP guides) may be utilised as part of the documentation, provided that those materials reflect the specific food operations of the business.

Documentation examples are:

- Hazard analysis;
- CCP determination;
- Critical limit determination.

Record examples are:

- CCP monitoring activities;
- Deviations and associated corrective actions;

- Verification procedures performed;
- Modifications to the HACCP plan.

An example of a HACCP worksheet for the development of a HACCP plan is attached as Diagram 3.

A simple record-keeping system can be effective and easily communicated to employees. It may be integrated into existing operations and may use existing paperwork, such as delivery invoices and checklists to record, for example, product temperatures.

TRAINING

Training of personnel in industry, government and academia in HACCP principles and applications and increasing awareness of consumers are essential elements for the effective implementation of HACCP. As an aid in developing specific training to support a HACCP plan, working instructions and procedures should be developed which define the tasks of the operating personnel to be stationed at each Critical Control Point.

Cooperation between primary producer, industry, trade groups, consumer organizations, and responsible authorities is of vital importance. Opportunities should be provided for the joint training of industry and control authorities to encourage and maintain a continuous dialogue and create a climate of understanding in the practical application of HACCP.

DIAGRAM 1
LOGIC SEQUENCE FOR THE APPLICATION OF HACCP

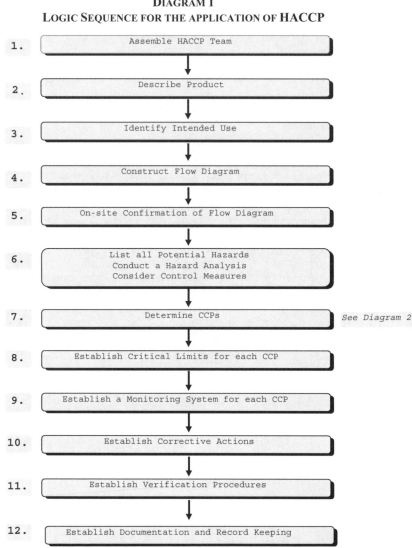

1. Assemble HACCP Team

2. Describe Product

3. Identify Intended Use

4. Construct Flow Diagram

5. On-site Confirmation of Flow Diagram

6. List all Potential Hazards
Conduct a Hazard Analysis
Consider Control Measures

7. Determine CCPs   *See Diagram 2*

8. Establish Critical Limits for each CCP

9. Establish a Monitoring System for each CCP

10. Establish Corrective Actions

11. Establish Verification Procedures

12. Establish Documentation and Record Keeping

DIAGRAM 2
EXAMPLE OF DECISION TREE TO IDENTIFY CCPS
(answer questions in sequence)

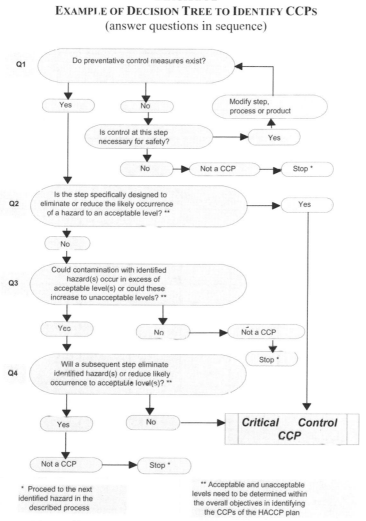

* Proceed to the next
identified hazard in the
described process

** Acceptable and unacceptable
levels need to be determined within
the overall objectives in identifying
the CCPs of the HACCP plan

DIAGRAM 3

EXAMPLE OF A HACCP WORKSHEET

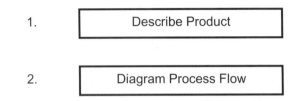

1. | Describe Product |

2. | Diagram Process Flow |

3.

List							
Step	Hazard(s)	Control Measure(s)	CCPs	Critical Limit(s)	Monitoring Procedure(s)	Corrective Action(s)	Record(s)

4. | Verification |

PRINCIPLES FOR THE ESTABLISHMENT AND APPLICATION OF MICROBIOLOGICAL CRITERIA FOR FOODS

CAC/GL 21 - 1997

INTRODUCTION

These Principles are intended to give guidance on the establishment and application of microbiological criteria for foods at any point in the food chain from primary production to final consumption.

The safety of foods is principally assured by control at the source, product design and process control, and the application of Good Hygienic Practices during production, processing (including labelling), handling, distribution, storage, sale, preparation and use, in conjunction with the application of the HACCP system. This preventive approach offers more control than microbiological testing because the effectiveness of microbiological examination to assess the safety of foods is limited. Guidance for the establishment of HACCP based systems is detailed in Hazard Analysis and Critical Control Point System and Guidelines for its Application (Annex to CAC/RCP 1-1969, Rev. 3-1997, Amd. 1999).

Microbiological criteria should be established according to these principles and be based on scientific analysis and advice, and, where sufficient data are available, a risk analysis appropriate to the foodstuff and its use. Microbiological criteria should be developed in a transparent fashion and meet the requirements of fair trade. They should be reviewed periodically for relevance with respect to emerging pathogens, changing technologies, and new understandings of science.

1. DEFINITION OF MICROBIOLOGICAL CRITERION

A microbiological criterion for food defines the acceptability of a product or a food lot, based on the absence or presence, or number of microorganisms including parasites, and/or quantity of their toxins/metabolites, per unit(s) of mass, volume, area or lot.

2. COMPONENTS OF MICROBIOLOGICAL CRITERIA FOR FOODS

A microbiological criterion consists of:

- a statement of the microorganisms of concern and/or their toxins/metabolites and the reason for that concern (see § 5.1);
- the analytical methods for their detection and/or quantification (see § 5.2);
- a plan defining the number of field samples to be taken and the size of the analytical unit (see § 6);
- microbiological limits considered appropriate to the food at the specified point(s) of the food chain (see § 5.3);
- the number of analytical units that should conform to these limits.

A microbiological criterion should also state:

- the food to which the criterion applies;
- the point(s) in the food chain where the criterion applies; and
- any actions to be taken when the criterion is not met.

When applying a microbiological criterion for assessing products, it is essential, in order to make the best use of money and manpower, that only appropriate tests be applied (see § 5) to those foods and at those points in the food chain that offer maximum benefit in providing the consumer with a food that is safe and suitable for consumption.

3. PURPOSES AND APPLICATION OF MICROBIOLOGICAL CRITERIA FOR FOODS

Microbiological criteria may be used to formulate design requirements and to indicate the required microbiological status of raw materials, ingredients and end-products at any stage of the food chain as appropriate. They may be relevant to the examination of foods, including raw materials and ingredients, of unknown or uncertain origin or when other means of verifying the efficacy of HACCP-based systems and Good Hygienic Practices are not available. Generally, microbiological criteria may be applied to define the distinction between acceptable and unacceptable raw materials, ingredients, products, lots, by regulatory authorities and/or food business operators. Microbiological criteria may also be used to determine that processes are consistent with *the General Principles of Food Hygiene*.

3.1.1 APPLICATION BY REGULATORY AUTHORITIES

Microbiological criteria can be used to define and check compliance with the microbiological requirements.

Mandatory microbiological criteria shall apply to those products and/or points of the food chain where no other more effective tools are available, and where they are expected to improve the degree of protection offered to the consumer. Where these are appropriate they shall be product-type specific and only applied at the point of the food chain as specified in the regulation.

In situations of non-compliance with microbiological criteria, depending on the assessment of the risk to the consumer, the point in the food chain and the product-type specified, the regulatory control actions may be sorting, reprocessing, rejection or destruction of product, and/or further investigation to determine appropriate actions to be taken.

3.1.2 APPLICATION BY A FOOD BUSINESS OPERATOR

In addition to checking compliance with regulatory provisions (see § 3.1.1) microbiological criteria may be applied by food business operators to formulate design requirements and to examine end-products as one of the measures to verify and/or validate the efficacy of the HACCP plan.

Such criteria will be specific for the product and the stage in the food chain at which they will apply. They may be stricter than the criteria used for regulatory purposes and should, as such, not be used for legal action.

Microbiological criteria are not normally suitable for monitoring Critical Limits as defined in *Hazard Analysis and Critical Control Point System and Guidelines for its Application* (Annex to CAC/RCP 1-1969, Rev. 3-1997). Monitoring procedures must be able to detect loss of control at a Critical Control Point (CCP). Monitoring should provide this information in time for corrective actions to be taken to regain control before there is a need to reject the product. Consequently, on-line measurements of physical and chemical parameters are often preferred to microbiological testing because results are often available more rapidly and at the production site. Moreover, the establishment of Critical Limits may need other considerations than those described in this document.

4. GENERAL CONSIDERATIONS CONCERNING PRINCIPLES FOR ESTABLISHING AND APPLYING MICROBIOLOGICAL CRITERIA

A microbiological criterion should be established and applied only where there is a definite need and where its application is practical. Such need is demonstrated, for example, by epidemiological evidence that the food under consideration may represent a public health risk and that a criterion is meaningful for consumer protection, or as the result of a risk assessment. The criterion should be technically attainable by applying Good Manufacturing Practices (Codes of Practice).

To fulfil the purposes of a microbiological criterion, consideration should be given to:

- the evidence of actual or potential hazards to health;
- the microbiological status of the raw material(s);
- the effect of processing on the microbiological status of the food;

- the likelihood and consequences of microbial contamination and/or growth during subsequent handling, storage and use;

- the category(s) of consumers concerned;

- the cost/benefit ratio associated with the application of the criterion; and

- the intended use of the food.

The number and size of analytical units per lot tested should be as stated in the sampling plan and should not be modified. However, a lot should not be subjected to repeated testing in order to bring the lot into compliance.

5. *MICROBIOLOGICAL ASPECTS OF CRITERIA*

5.1 MICROORGANISMS, PARASITES AND THEIR TOXINS/METABOLITES OF IMPORTANCE IN A PARTICULAR FOOD

For the purpose of this document these include:

- bacteria, viruses, yeasts, moulds, and algae;

- parasitic protozoa and helminths;

- their toxins/metabolites.

The microorganisms included in a criterion should be widely accepted as relevant - as pathogens, as indicator organisms or as spoilage organisms - to the particular food and technology. Organisms whose significance in the specified food is doubtful should not be included in a criterion.

The mere finding, with a presence-absence test, of certain organisms known to cause foodborne illness (e.g. *Clostridium perfringens, Staphylococcus aureus* and *Vibrio parahaemolyticus*) does not necessarily indicate a threat to public health.

Where pathogens can be detected directly and reliably, consideration should be given to testing for them in preference to testing for indicator organisms. If a test for an indicator organism is applied, there should be a clear statement whether the test is used to indicate unsatisfactory hygienic practices or a health hazard.

5.2 MICROBIOLOGICAL METHODS

Whenever possible, only methods for which the reliability (accuracy, reproducibility, inter- and intra-laboratory variation) has been statistically established in comparative or collaborative studies in several laboratories should be used. Moreover, preference should be given to methods which have been validated for the commodity concerned preferably in relation to reference methods elaborated by international organizations. While methods should be the most sensitive and reproducible for the purpose, methods to be used for in-plant testing might often sacrifice to some degree sensitivity and reproducibility in the interest of speed and simplicity. They should, however, have been proved to give a sufficiently reliable estimate of the information needed.

Methods used to determine the suitability for consumption of highly perishable foods, or foods with a short shelf-life, should be chosen wherever possible so that the results of microbiological examinations are available before the foods are consumed or exceed their shelf-life.

The microbiological methods specified should be reasonable with regard to complexity, availability of media, equipment etc., ease of interpretation, time required and costs.

5.3 MICROBIOLOGICAL LIMITS

Limits used in criteria should be based on microbiological data appropriate to the food and should be applicable to a variety of similar products. They should therefore be based on data gathered at various production establishments operating under Good Hygienic Practices and applying the HACCP system.

In the establishment of microbiological limits, any changes in the microflora likely to occur during storage and distribution (e.g. decrease or increase in numbers) should be taken into account.

Microbiological limits should take into consideration the risk associated with the microorganisms, and the conditions under which the food is expected to be handled and consumed. Microbiological limits should also take account of the likelihood of uneven distribution of microorganisms in the food and the inherent variability of the analytical procedure.

If a criterion requires the absence of a particular microorganism, the size and number of the analytical unit (as well as the number of analytical sample units) should be indicated.

6.   *SAMPLING PLANS, METHODS AND HANDLING*

A sampling plan includes the sampling procedure and the decision criteria to be applied to a lot, based on examination of a prescribed number of sample units and subsequent analytical units of a stated size by defined methods. A well-designed sampling plan defines the probability of detecting microorganisms in a lot, but it should be borne in mind that no sampling plan can ensure the absence of a particular organism. Sampling plans should be administratively and economically feasible.

In particular, the choice of sampling plans should take into account:

- risks to public health associated with the hazard;
- the susceptibility of the target group of consumers;
- the heterogeneity of distribution of microorganisms where variables sampling plans are employed; and
- the Acceptable Quality Level[3] and the desired statistical probability of accepting a non-conforming lot.

For many applications, 2-or 3-class attribute plans may prove useful. [4]

The statistical performance characteristics or operating characteristics curve should be provided in the sampling plan. Performance characteristics provide specific information to estimate the probability of accepting a non-conforming lot. The sampling method should be defined in the sampling plan. The time between taking the field samples and analysis should be as short as reasonably possible, and during transport to the laboratory the conditions (e.g. temperature) should not allow increase or decrease of the numbers of the target organism, so that the results reflect - within the limitations given by the sampling plan - the microbiological conditions of the lot.

---

[3] The Acceptable Quality Level (AQL) is the percentage of non-conforming sample units in the entire lot for which the sampling plan will indicate lot acceptance for a prescribed probability (usually 95 per cent).

[4] See ICMSF: Microorganisms in Foods, 2. Sampling for Microbiological Analysis. Principles and Specific Applications, 2nd Edition, Blackwell Scientific Publications, 1986 (ISBN-0632-015-675).

7.   REPORTING

The test report shall give the information needed for complete identification of the sample, the sampling plan, the test method, the results and, if appropriate, their interpretation.

PRINCIPLES AND GUIDELINES FOR THE CONDUCT OF MICROBIOLOGICAL RISK ASSESSMENT

CAC/GL-30 (1999)

INTRODUCTION

Risks from microbiological hazards are of immediate and serious concern to human health. Microbiological risk analysis is a process consisting of three components: Risk assessment, risk management, and risk communication, which has the overall objective to ensure public health protection. This document deals with risk assessment which is a key element in assuring that sound science is used to establish standards, guidelines and other recommendations for food safety to enhance consumer protection and facilitate international trade. The microbiological risk assessment process should include quantitative information to the greatest extent possible in the estimation of risk. A microbiological risk assessment should be conducted using a structured approach such as that described in this document. This document will be of primary interest to governments although other organizations, companies, and other interested parties who need to prepare a microbiological risk assessment will find it valuable. Since microbiological risk assessment is a developing science, implementation of these guidelines may require a period of time and may also require specialized training in the countries that consider it necessary. This may be particularly the case for developing countries. Although microbiological risk assessment is the primary focus of this document, the method can also be applied to certain other classes of biological hazards.

1.   SCOPE

The scope of this document applies to risk assessment of microbiological hazards in food.

2.   DEFINITIONS

The definitions cited here are to facilitate the understanding of certain words or phrases used in this document.

Where available the definitions are those adopted for microbiological, chemical, or physical agents, risk management and risk communication on an interim basis at the 22nd Session of the Codex Alimentarius Commission. The CAC adopted these definitions on an interim basis because they are subject to modification in the light of developments in the science of risk analysis and as a result of efforts to harmonize similar definitions across various disciplines.

Dose-Response Assessment - The determination of the relationship between the magnitude of exposure (dose) to a chemical, biological or physical agent and the severity and/or frequency of associated adverse health effects (response).

Exposure Assessment - The qualitative and/or quantitative evaluation of the likely intake of biological, chemical, and physical agents via food as well as exposures from other sources if relevant.

Hazard - A biological, chemical or physical agent in, or condition of, food with the potential to cause an adverse health effect.

Hazard Characterization - The qualitative and/or quantitative evaluation of the nature of the adverse health effects associated with the hazard. For the purpose of microbiological risk assessment the concerns relate to microorganisms and/or their toxins.

Hazard Identification - The identification of biological, chemical, and physical agents capable of causing adverse health effects and which may be present in a particular food or group of foods.

Quantitative Risk Assessment - A risk assessment that provides numerical expressions of risk and indication of the attendant uncertainties (stated in the 1995 Expert Consultation definition on Risk Analysis).

Qualitative Risk Assessment - A risk assessment based on data which, while forming an inadequate basis for numerical risk estimations, nonetheless, when conditioned by prior expert knowledge and identification of attendant uncertainties permits risk ranking or separation into descriptive categories of risk.

Risk - A function of the probability of an adverse health effect and the severity of that effect, consequential to a hazard(s) in food.

Risk Analysis - A process consisting of three components: Risk assessment, risk management and risk communication.

Risk Assessment - A scientifically based process consisting of the following steps: (i) hazard identification, (ii) hazard characterization, (iii) exposure assessment, and (iv) risk characterization.

Risk Characterization - The process of determining the qualitative and/or quantitative estimation, including attendant uncertainties, of the probability of occurrence and severity of known or potential adverse health effects in a given population based on hazard identification, hazard characterization and exposure assessment.

Risk Communication - The interactive exchange of information and opinions concerning risk and risk management among risk assessors, risk managers, consumers and other interested parties.

Risk Estimate - Output of risk characterization.

Risk Management - The process of weighing policy alternatives in the light of the results of risk assessment and, if required, selecting and implementing appropriate control[5] options, including regulatory measures.

---

[5]   Control means prevention, elimination, or reduction of hazards and/or minimization of risks.

Sensitivity analysis - A method used to examine the behavior of a model by measuring the variation in its outputs resulting from changes to its inputs.

Transparent - Characteristics of a process where the rationale, the logic of development, constraints, assumptions, value judgements, decisions, limitations and uncertainties of the expressed determination are fully and systematically stated, documented, and accessible for review.

Uncertainty analysis - A method used to estimate the uncertainty associated with model inputs, assumptions and structure/form.

3. GENERAL PRINCIPLES OF MICROBIOLOGICAL RISK ASSESSMENT

1. Microbiological risk assessment should be soundly based upon science.

2. There should be a functional separation between risk assessment and risk management.

3. Microbiological risk assessment should be conducted according to a structured approach that includes hazard identification, hazard characterization, exposure assessment, and risk characterization.

4. A microbiological risk assessment should clearly state the purpose of the exercise, including the form of risk estimate that will be the output.

5. The conduct of a microbiological risk assessment should be transparent.

6. Any constraints that impact on the risk assessment such as cost, resources or time, should be identified and their possible consequences described.

7. The risk estimate should contain a description of uncertainty and where the uncertainty arose during the risk assessment process.

8. Data should be such that uncertainty in the risk estimate can be determined; data and data collection systems should, as far as possible, be of sufficient quality and precision that uncertainty in the risk estimate is minimized.

9. A microbiological risk assessment should explicitly consider the dynamics of microbiological growth, survival, and death in foods and the complexity of the interaction (including sequelae) between human and agent following consumption as well as the potential for further spread.

10. Wherever possible, risk estimates should be reassessed over time by comparison with independent human illness data.

11. A microbiological risk assessment may need reevaluation, as new relevant information becomes available.

4.   GUIDELINES FOR APPLICATION

These Guidelines provide an outline of the elements of a Microbiological Risk Assessment indicating the types of decisions that need to be considered at each step.

4.1   GENERAL CONSIDERATIONS

The elements of risk analysis are: Risk assessment, risk management, and risk communication.   The functional separation of risk assessment from risk management helps assure that the risk assessment process is unbiased. However, certain interactions are needed for a comprehensive and systematic risk assessment process. These may include ranking of hazards and risk assessment policy decisions.   Where risk management issues are taken into account in risk assessment, the decision-making process should be transparent. It is the transparent unbiased nature of the process that is important, not who is the assessor or who is the manager.

Whenever practical, efforts should be made to provide a risk assessment process that allows contributions by interested parties. Contributions by interested parties in the risk assessment process can improve the transparency of the risk assessment, increase the quality of risk assessments through additional expertise and information, and facilitate risk communication by increasing the credibility and acceptance of the results of the risk assessment.

Scientific evidence may be limited, incomplete or conflicting. In such cases, transparent informed decisions will have to be made on how to complete the risk assessment process.   The importance of using high quality information when conducting a risk assessment is to reduce uncertainty and to increase the reliability of the risk estimate. The use of quantitative information is encouraged to the extent possible, but the value and utility of qualitative information should not be discounted.

It should be recognized that sufficient resources will not always be available and constraints are likely to be imposed on the risk assessment that will influence the quality of the risk estimate. Where such resource constraints apply, it is important for transparency purposes that these constraints be described in the formal record. Where appropriate, the record should include an evaluation of the impact of the resource constraints on the risk assessment.

4.2   STATEMENT OF PURPOSE OF RISK ASSESSMENT

At the beginning of the work the specific purpose of the particular risk assessment being carried out should be clearly stated. The output form and possible output alternatives of the risk assessment should be defined. Output might, for example, take the form of an estimate of the prevalence of illness, or an estimate of annual rate (incidence of human illness per 100,000) or an estimate of the rate of human illness and severity per eating occurrence.

The microbiological risk assessment may require a preliminary investigation phase. In this phase, evidence to support farm-to-table modelling of risk might be structured or mapped into the framework of risk assessment.

4.3   HAZARD IDENTIFICATION

For microbial agents, the purpose of hazard identification is to identify the microorganisms or the microbial toxins of concern with food. Hazard identification will predominately be a qualitative process. Hazards can be identified from relevant data sources. Information on hazards can be obtained from scientific literature, from databases such as those in the food industry, government agencies, and relevant international organizations and through solicitation of opinions of experts.   Relevant information includes data in areas such as: clinical studies, epidemiological studies and surveillance, laboratory animal studies, investigations of the characteristics of microorganisms, the interaction between microorganisms and their environment through the food chain from primary production up to and including consumption, and studies on analogous microorganisms and situations.

4.4   EXPOSURE ASSESSMENT

Exposure assessment includes an assessment of the extent of actual or anticipated human exposure. For microbiological agents, exposure assessments might be based on the potential extent of food contamination by a particular agent or its toxins, and on dietary information. Exposure assessment should specify the unit of food that is of interest, i.e., the portion size in most/all cases of acute illness.

Factors that must be considered for exposure assessment include the frequency of contamination of foods by the pathogenic agent and its level in those foods over time. For example, these factors are influenced by the characteristics of the pathogenic agent, the microbiological ecology of the food, the initial contamination of the raw material including considerations of regional differences and seasonality of production, the level of sanitation and process controls, the methods of processing, packaging, distribution and storage of the foods, as well as any preparation steps such as cooking and holding. Another factor that must be considered in the assessment is patterns of consumption. This relates to socio-economic and cultural backgrounds, ethnicity, seasonality, age differences (population demographics), regional differences, and consumer preferences and behavior. Other factors to be considered include: the role of the food handler as a source of contamination, the amount of hand contact with the product, and the potential impact of abusive environmental time/temperature relationships.

Microbial pathogen levels can be dynamic and while they may be kept low, for example, by proper time/temperature controls during food processing, they can substantially increase with abuse conditions (for example, improper food storage temperatures or cross contamination from other foods). Therefore, the exposure assessment should describe the pathway from production to consumption. Scenarios can be constructed to predict the range of possible exposures. The scenarios might reflect effects of processing, such as hygienic design, cleaning and disinfection, as well as the time/temperature and other conditions of the food history, food handling and consumption patterns, regulatory controls, and surveillance systems.

Exposure assessment estimates the level, within various levels of uncertainty, of microbiological pathogens or microbiological toxins, and the likelihood of their occurrence in foods at the time of consumption. Qualitatively foods can be categorized according to the likelihood that the foodstuff will or will not be contaminated at its source; whether or not the food can support the growth of the pathogen of concern; whether there is substantial potential for abusive handling of the food; or whether the food will be subjected to a heat process. The presence, growth, survival, or death of microorganisms, including pathogens in foods, are influenced by processing and packaging, the storage environment, including the temperature of storage, the relative humidity of the environment, and the gaseous composition of the atmosphere. Other relevant factors include pH, moisture content or water activity (a_w), nutrient content, the presence of antimicrobial substances, and competing microflora. Predictive microbiology can be a useful tool in an exposure assessment.

4.5   HAZARD CHARACTERIZATION

This step provides a qualitative or quantitative description of the severity and duration of adverse effects that may result from the ingestion of a microorganism or its toxin in food. A dose-response assessment should be performed if the data are obtainable.

There are several important factors that need to be considered in hazard characterization. These are related to both the microorganism, and the human host. In relation to the microorganism the following are important: microorganisms are capable of replicating; the virulence and infectivity of microorganisms can change depending on their interaction with the host and the environment; genetic material can be transferred between microorganisms leading to the transfer of characteristics such as antibiotic resistance and virulence factors; microorganisms can be spread through secondary and tertiary transmission; the onset of clinical symptoms can be substantially delayed following exposure; microorganisms can persist in certain individuals leading to continued excretion of the microorganism and continued risk of spread of infection; low doses of some microorganisms can in some cases cause a severe effect; and the attributes of a food that may alter the microbial pathogenicity, e.g., high fat content of a food vehicle.

In relation to the host the following may be important: genetic factors such as human leucocyte antigen (HLA) type; increased susceptibility due to breakdowns of physiological barriers; individual host susceptibility characteristics such as age, pregnancy, nutrition, health and medication status, concurrent infections, immune status and previous exposure history; population characteristics such as population immunity, access to and use of medical care, and persistence of the organism in the population.

A desirable feature of hazard characterization is ideally establishing a dose-response relationship. When establishing a dose-response relationship, the different end points, such as infection or illness, should be taken into consideration. In the absence of a known dose-response relationship, risk assessment tools such as expert elicitations could be used to consider various factors, such as infectivity, necessary to describe hazard characterizations. Additionally, experts may be able to devise ranking systems so that they can be used to characterize severity and/or duration of disease.

4.6   RISK CHARACTERIZATION

Risk characterization represents the integration of the hazard identification, hazard characterization, and exposure assessment determinations to obtain a risk estimate; providing a qualitative or quantitative estimate of the likelihood and severity of the adverse effects which could occur in a given population, including a description of the uncertainties associated with these estimates.

These estimates can be assessed by comparison with independent epidemiological data that relate hazards to disease prevalence.

Risk characterization brings together all of the qualitative or quantitative information of the previous steps to provide a soundly based estimate of risk for a given population. Risk characterization depends on available data and expert judgements. The weight of evidence integrating quantitative and qualitative data may permit only a qualitative estimate of risk.

The degree of confidence in the final estimation of risk will depend on the variability, uncertainty, and assumptions identified in all previous steps. Differentiation of uncertainty and variability is important in subsequent selections of risk management options. Uncertainty is associated with the data themselves, and with the choice of model. Data uncertainties include those that might arise in the evaluation and extrapolation of information obtained from epidemiological, microbiological, and laboratory animal studies. Uncertainties arise whenever attempts are made to use data concerning the occurrence of certain phenomena obtained under one set of conditions to make estimations or predictions about phenomena likely to occur under other sets of conditions for which data are not available. Biological variation includes the differences in virulence that exist in microbiological populations and variability in susceptibility within the human population and particular subpopulations.

It is important to demonstrate the influence of the estimates and assumptions used in risk assessment; for quantitative risk assessment this can be done using sensitivity and uncertainty analyses.

4.7 DOCUMENTATION

The risk assessment should be fully and systematically documented and communicated to the risk manager. Understanding any limitations that influenced a risk assessment is essential for transparency of the process that is important in decision making. For example, expert judgements should be identified and their rationale explained. To ensure a transparent risk assessment a formal record, including a summary, should be prepared and made available to interested independent parties so that other risk assessors can repeat and critique the work. The formal record and summary should indicate any constraints, uncertainties, and assumptions and their impact on the risk assessment.

4.8 REASSESSMENT

Surveillance programs can provide an ongoing opportunity to reassess the public health risks associated with pathogens in foods as new relevant information and data become available. Microbiological risk assessors may have the opportunity to compare the predicted risk estimate from microbiological risk assessment models with reported human illness data for the purpose of gauging the reliability of the predicted estimate. This comparison emphasizes the iterative nature of modelling. When new data become available, a microbiological risk assessment may need to be revisited.

PUBLICATION HISTORY

This booklet is an extract of Volume 1B - **General Requirements (Food Hygiene)** of the *Codex Alimentarius*. The following table indicates previous versions of these texts and the reference to the draft texts prepared by the Codex Committee on Food Hygiene.

DOCUMENT	REFERENCES
Recommended International Code of Practice - General Principles of Food Hygiene:	CAC/RCP-1 (1969)
Revision 1	1979
Revision 2	1985
Revision 3 (Current)	1997
Draft adopted by the 22nd Session of the Commission	ALINORM 97/13, Appendix II
Amendments regarding rinsing adopted by the 23rd Session of the Commission	ALINORM 99/13A, Appendix III
Hazard Analysis and Critical Control Point (HACCP) System and Guidelines for its Application	CAC/GL 18-1993
Revision 1	Annex to CAC/RCP-1 (1969), Rev. 3 (1997)
Prior draft	ALINORM 93/13A, Appendix II
Draft adopted by the 22nd Session of the Commission	ALINORM 97/13A, Appendix II
Revision (Current)	Annex to CAC/RCP-1 (1969), Rev.3 (1997)
Draft adopted by the 26th Session of the Commission	ALINORM 03/13A, Appendix II

Principles for the Establishment and Application of Microbiological Criteria for Foods	Published in the Procedural Manual of the Codex Alimentarius Commission, Sixth to Ninth Editions (1986-1995)
Revision 1 (Current)	CAC/GL-21 (1997)
Draft adopted by the 22nd Session of the Commission	ALINORM 97/13A, Appendix III
Principles and Guidelines for the Conduct of Microbiological Risk Assessment	CAC/GL 30-1999
Draft adopted by the 23rd Session of the Commission	ALINORM 99/13A, Appendix II

INDEX

Definition, 33
Virulence, 58, 59
Viruses, 47

W

Water activity (a$_w$), 57

Y

Yeasts, 47